AUDREY KIBAMBA DE BOUANSA

CLEANING UP THE HUMAN SKELETON AND GENOME

AUDREY KIBAMBA DE BOUANSA

CLEANING UP THE HUMAN SKELETON AND GENOME

USING THE DISCRIMINANT METHOD IN A GRAND-CARRE

ScienciaScripts

Imprint

Any brand names and product names mentioned in this book are subject to trademark, brand or patent protection and are trademarks or registered trademarks of their respective holders. The use of brand names, product names, common names, trade names, product descriptions etc. even without a particular marking in this work is in no way to be construed to mean that such names may be regarded as unrestricted in respect of trademark and brand protection legislation and could thus be used by anyone.

Cover image: www.ingimage.com

This book is a translation from the original published under ISBN 978-620-6-71620-4.

Publisher:
Sciencia Scripts
is a trademark of
Dodo Books Indian Ocean Ltd. and OmniScriptum S.R.L publishing group

120 High Road, East Finchley, London, N2 9ED, United Kingdom
Str. Armeneasca 28/1, office 1, Chisinau MD-2012, Republic of Moldova, Europe
Printed at: see last page
ISBN: 978-620-7-92169-0

CLEANING THE HUMAN SKELETON AND GENOME USING THE DISCRIMINANT METHOD IN A GRAND-CARRE

AUDREY KIBAMBA DE BOUANSA GARE DE LUMIERE

THESIS

Research in Mathematics of Medical Engineering

Specialty: Cleansing the Human Skeleton and Genome with
Natural Light and Digital Pressure

Sciences related to Spirituality and African Art

Realized

Visit

Laboratoire Universitaire de Contrôle, Correction, Complément et
Création des Théories Scientifiques (LUC4) at the Centre de
Géométrie Absolue (CGA) in Brazzaville

By Audrey KIBAMBA MOUNTOU

Audrey KIBAMBA DE BOUANSA GARE DE LUMIERE

With the collaboration of Professor Hassan BAKHSSIS,
Mathematician, Writer, decorated with the rank of Chevalier in
the Order of Academic Palms in France.

And

Paul YOTCHO, Fifth-Rank Teacher, University Professor,
Mathematician, Physicist, Researcher, Writer.

Contents

Preface 4

Introduction 7

Chapter I: Discovery of Discriminants in the composition of the human body 11

Chapter II: What is the Discriminant Method? 15

Chapter III: What causes of human skeletal and genomic dysfunction will the Discriminant Method resolve? 18

Chapter IV: The Different Types of Balls that Cause Skeletal and Human Genome Dysfunction 20

Chapter V: Energy variations in the human body: a global public health issue 23

Chapter VI: The world of medical engineering 26

Conclusion 29

Afterword 33

References 35

Foreword

"It's a thesis traversed by scientific and technical themes in which the indispensable irrationality interacts: African Art and Spirituality (an element of stability in any science so that the latter doesn't cross the boundaries of ethics, morality... I reread this precious thesis with great interest, during which it's not so easy to tackle all the themes, analyses and reflections. The skilful and ingenious researcher Audrey KIBAMBA had a head start, for he constructed his thesis in such a way that all the themes dealt with converged like guidelines towards the same vanishing point. I'm going to focus my observations on this vanishing point: knowledge must encompass Human Society without distinction!

In biology, bacteria are known to cooperate, exchange information, evolve and even advance their ability to evolve. Ants do the same. Human beings need to do the same too, to evolve better in this fast-paced, dizzying world. This world is progressing spectacularly in many areas, and we are experiencing a major transition in our evolution: the astounding discoveries of artificial intelligence and genetic discoveries are posing unprecedented challenges to the human species. Artificial intelligence is having an impact in all fields of knowledge, but with major concerns about ethics, morality and the replacement of man by the "intelligent" robot.

Professor Audrey KIBAMBA's thesis answers this dizzying question about science when it is not irrigated by the sacredness and solid irrationality of the breaths dear to our ancestors. This thesis seems to me to be solid in several respects, as it brings to light other elements that prevent the citizens of one part of the planet from developing and flourishing, and also from selling their knowledge! As far as I'm concerned, this thesis is like a manifesto for well-trained but free minds! A research thesis that is both scientific and reflective. It takes a cultural anthropological approach. The explanations are clearly indicated and the approach adopted is rigorously scientific. And as Edgar Morin points out, the human being is trinitarian, defined in a three-term loop of species/individual/society, where each term is necessary to the existence of the others. And as Professor Audrey KIBAMBA implicitly states in her thesis: society is inside the individual, inscribed in his culture, language, mores and sacredness. And in any science, recognizing the demands of reality does not mean ignoring the demands of another order. Contemporary society generally highlights exclusively objective knowledge with measurable, profitable or even mercantile effects. This thesis introduces another irrational dimension, centered on the human above all else: an expanded unconscious as a concept.

André Malraux was an infallible advocate of culture: "It is in culture that the heart of a civilization beats", he said...

Professor Audrey KIBAMBA is attempting to open up a neglected breach in scientific analysis, so that it can be global, taking into account the evolution and also the slowness of the five continents. Its aim is to access a better future, to think and build perspectives capable of contributing to the collective and liberating progress of all peoples and societies without distinction. "The human genome has been completely deciphered. And "that a genetic writing of the history of human populations is underway." (Source la science au présent 2024 Encyclopédie Universalise).

Artificial intelligence, functional ecology, genomics... science and technology are constantly lifting the veils that previously covered unsuspected areas of knowledge about life, matter, the Earth or our Universe, often calling into question well-established certainties. As researcher Audrey KIBAMBA points out in this vanishing point.

Scientific discoveries also sketch out imaginable futures, more or less desirable or worrying depending on what we do with our knowledge and technologies. Professor Audrey KIBAMBA's thesis corroborates this idea, reported in the book Science au présent 2024. So many points of convergence with the analyses of the author of this thesis. The latter is a step forward in the construction of universal, shared knowledge. Every year, fortuitous observations such as those made by Professor KIBAMBA, natural phenomena or the results of significant research, supplement or challenge previous knowledge. The author insists on ethics and sacredness (La Lumière), whose sources emanate from the five continents, but are universal where no nation is servile. A science of consciousness, of belonging with dignity to the human species.

I would like to end by thanking Professor Audrey KIBAMBA for the quality of her research and her concern to think first and foremost of others... in order to achieve an earthy, just and united citizenship."

Professor Hassan BAKHSISS

Introduction

Humanity's elders have had the merit of advancing humanity through the power of their revolutionary thinking, which reverberates across the screen of space-time. They were true intellectuals of head and heart, with an incredible genius for know-how, which they accompanied with know-how in their combat and in their ideal of always thinking well, rightly and doing great things for others, in the light of the wisdom of love, which always places man at the center of the general interest.

But they were also wrong to misdefine certain concepts in their time, which are now blocking humanity in an apostasy of knowledge in which we have become partisans of the least effort, acting like veritable robots whose mission is to memorize only what the elders of humanity have said, as the gospel that we should swallow without a second thought, or think in a new frame of reference to redefine things in the battle of our times, which is the essence and urgency of meeting the challenges of the millennium through research, innovation and discovery.

Personally, I take issue with the notion that humanity's elders were the ones who solved all of humanity's problems, and that we shouldn't be thinking, reflecting or even reasoning about them in this day and age (there's nothing new under the sun). This is a big mistake. Because intelligence escapes from one generation to the next in the foliage of time.

And human history is made up of great challenges and intergenerational fire in the universal light of reason. The elders were able to say what they believed to be true and right in their time, which is naturally different from ours. Realities are no longer the same. Life itself is dynamic, not static. Things are no longer the same. They evolve from one era to the next.

So, life is not a scalar. Everything moves and dies in a dialectic of spirit light and matter in motion of the Discriminants of the second-degree trinomial in a twelve-dimensional space-time of etheric sub-quantum fields.

Mankind's elders did not and could not have had a monopoly on knowledge and learning, let alone the intellectual capacity to solve all of humanity's problems definitively in their own time. They were also limited in the way they saw, understood, interpreted and defined certain things or concepts, despite the power of their revolutionary thinking, the effects of which we have inherited in the eternal memory of generations.

They are at the origin of this apostasy of knowledge, which obviously traps mankind in dogmas and doctrines that are, in reality, a prison for the soul and, above all, for the development of the spirit, which needs to be deconstructed and rebuilt in a new frame of reference that takes account of our current realities in people's lives. For every generation is a child of its time, in the universal light of reason.

We don't need to write pages to state a scientific truth drawn from experimental soil on the basis of intuition and mathematical astuteness, and its demonstration by a refinement of thought.

Each discovery is humanity sharing with others in the culture of the search for excellence in continuity and discontinuity. That's why you can't be a Master or a Doctor without a thesis of genuine research into the universality of the light of reason through a refinement of thought.

Research, innovation and discovery are a world of contributions and achievements to move humanity forward through the power of mathematical thinking, which is reflected on the screen of space-time in intellectual production in self-education.

Man is a universe contained within himself, which he discovers when he seeks to penetrate the secrets and mysteries of nature by refining his mathematical thinking.

The human body is a living pyramid in its mathematical composition. A universe of knowledge and a vast site of ongoing research. Because we don't have all the information we need about the human body. It's a dynamic that takes into account the environment in which we find ourselves, and the influences that often modify our behavior, morphology and even anatomy.

The human body is transcendent and ascendant: it's a mystery that also needs to be tamed in the mathematics of medical engineering in a new specialty called Human Skeletal and Genome Cleansing by natural light and digital pressure in a Great Square of the Sacred Triangle of Intelligence containing two vertices. This is where the light of the spirit travels in dimension twelve, where the spirit is naturally linked to the body to form one.

The global health system is experiencing shortcomings and limitations when it comes to understanding the human body and its many global problems, and we need to find effective solutions to improve this system with the latest achievements and research contributions in medical engineering mathematics.

This is why the introduction of the Discriminant Method into the global health system on a planetary scale is a great cultural and technoscientific revolution to overcome some of the system's limitations and flaws in the marriage of research and innovation leading to the progress of nations.

A well-shaped head is the best universal diploma there is in the contract of competence, where man contributes his know-how, which he naturally accompanies with know-how to advance humanity in intellectual coherence and in the culture of the search for excellence as the search for true truth drawn from experimental soil.

We live in a world of achievements and contributions, in which the intelligences of men escape from each other in the foliage of time. A world that does not forgive those who want to live haphazardly in a century without finding something to nourish it. So, we are in a world of research and innovation leading to development on the basis of intellectual production in self-education.

The discovery of the Discriminant and its application in a Grand-Carré for cleaning the human skeleton and genome using natural light and digital pressure is a great revolution in the global health system at the dawn of the third millennium, to meet the challenges of our time through research, innovation and discovery.

The greatness of any spirit of light is measured by what it contributes to the progression of humanity on the screen of space-time. That's why there are intellectuals and true intellectuals, as well as professors and great professors, on the basis of their achievements and contributions in the fields of science and literature.

The future is not always the future or tomorrow, but also the present, which is being built now, on a slope of sweat, perseverance and the prudence of time, to build great victories in the culture of striving for excellence in people's skills.

The innovation that led us to write this research thesis in mathematics of medical engineering is the discovery of Discriminants in the composition of the architectural framework of the human body, in order to go beyond and complement the work of the theses of the ancient Greek scientists who believed, rightly or wrongly, that the human body is composed solely of atoms and molecules.

This thesis of scientific research clearly shows that what remains to be discovered is greater than what is known. For the human body is a vast site of permanent research and discovery in the universality of the light of reason through the refinement of mathematical thinking.

Chapter I: Discovery of Discriminants in the composition of the human body

My contributions to research and discovery are the concentrated history of solar humanity and the platform of its experiences, where to exist is to be outside oneself towards another than oneself in the universality of the light of reason.

My research work in Mathematics of Medical Engineering has led to a major discovery: the irrefutable presence of Discriminants in the architectural composition of the human body.

In Mathematics of Medical Engineering, my discovery is humanity sharing with others in the foliage of time. Research, innovation and discovery are the challenges to be met, where the spirit of light reveals itself in the struggle of its time to mark history in the memory of generations with the weapons of intelligence.

Each generation has its eternal fire, its inventive and creative geniuses, the struggle of its time, and its frame of reference that enables it to see and understand the reality of intelligible universes with the spectacles of pure thought, capable of explaining and interpreting it in the totality of knowledge and in the knowledge of the totality.

It's the dialectic of the light of the spirit and matter in motion of surveying, where the universe is a whole of knowledge and knowing, expressed by the Poetry of Scientific Imagery as the matrix of all sciences in interdisciplinarity, multidisciplinarity and transdisciplinarity, essential assets for accessing true knowledge on the architecture of true truth drawn from experimental soil.

The history of mankind is made up of great challenges, research and innovation, invention and creation of the mind, even discovery, where the brilliant minds of each generation tame nature by obeying it to penetrate its secrets and mysteries in a permanent dialogue based on intuition and mathematical astuteness through a refinement of thought.

Research or discovery is not a scalar. Rather, it is a dynamic, a static, which takes its evolution from the universality of the light of reason, which leads us into a marriage of innovation and the creation of the mind.

Man is a universe contained within himself, whose secrets and mysteries must be penetrated by the eternal fire of generations. It's a rational function of sums in steps of n-i factors of powerful determinants where any positive integer that adds up as many times as itself equals its second power.

So man is a living, spiritual pyramid, made up of atoms, molecules and discriminants that form the backbone of his architecture of reality, whose existence is of mathematical essence, as the roots of humanity.

Man is a completed construction of the Mathematics of Medical Engineering, which constitutes the language of the universe contained within him, to be transmitted to others. In seeking to understand the universe, man discovers himself as an isolated system whose energies transform into one another. He is an explorer and user of alchemy, combining its basic elements in the operative possibility of the operations of Arithmetic. It is in this perfect knowledge that man becomes master and possessor of nature, obeying it in the rites and customs of the light of spirit, which he fuses with matter to form a single entity in a twelve-dimensional space-time.

There, man experiences the fusion of two cones of light and darkness, leading to the discovery of the wisdom of the spirit through the revelation of the Zebra and Refracted Lightwave, based on intuition and mathematical astuteness, in a permanent dialogue with nature, whose spirituality is the only bridge that can exist between science and man. He becomes a researcher and a beacon of innovation, creativity and discovery for the progress of science in time and space.

In medical engineering mathematics, the Discriminant is the square of the difference between two numbers that are solutions of the second-degree trinomial with one unknown.

The human body is a mosaic of quantitative, qualitative and differential discriminants, as Audrey KIBAMBA and Omer FOUETOLO put it. This means that the human body is not made up of atoms and molecules alone, as Democritus, Leucippus, Hippocrates and many others said in ancient times. But also of Discriminants, as we have just demonstrated in this new millennium specialty for cleaning the human skeleton and genome by natural light and digital pressure, naturally adjoining the fluid mechanisms in a Grand-Carré.

Hence $F = p / s$.

Discriminants in the human body vary in size. They vary from one organ to another. The discriminant is an algebraic entity that we use to grid the human body into geometric entities that are nothing more than geometric squares of varying size, depending on the size of the organ to be cleaned.

The fundamental principle of the Discriminant is to circulate oxygen in the human body by means of natural light and digital pressure.

In reality, Discriminants are also conductors in their role of circulating red and white blood, oxygen and water to feed the entire body system.

This highly remarkable contribution to the progress of science proves that research, innovation and discovery are becoming essential in the new development policies of nations in the third millennium, where knowledge and know-how are key variables in the growth of free nations.

This research thesis in Mathematics of Medical Engineering is a great revolution in the world health system with the introduction of the Discriminant Method for cleansing the human skeleton and genome by natural light and digital pressure in a Grand-Square of the Sacred Triangle of Intelligence containing two transcendent and ascending vertices. Where the light of the spirit travels.

Chapter II: What is the Discriminant Method?

The Discriminant Method is therefore a resuscitator par excellence of human life, directly using aerobic oxygen (the oxygen in natural air), through the pressure of digital light applied to the surface of an organ.

It's sacred geometry applied to the surface of an organ squared into small geometric squares, with the rational intelligence of the neurons of touch to cleanse it, the fundamental principle of which is that of circulating oxygen in the human body through natural light and digital pressure in a Grand-Carré.

This method, known as the Discriminant, enables us to cleanse the nerves, kidneys, muscles, red and white blood, brain, knee, water and many others.

The Discriminant method is our life insurance policy for rejuvenating our organism by cleansing the skeleton and the human genome using natural light and digital pressure in a Grand-Carré.

Discriminants are naturally connected to the pores of the human body, enabling the human skeleton to breathe and function properly by receiving sufficient oxygen from the various pores of the human body.

They regulate the human body's respiratory system, using natural light and finger pressure. Discriminants are therefore links in the food chain of the human skeleton and genome.

When a Discriminant is clogged, the blood in that area is left behind. However, the organ creates a rescue mechanism on the surface of the muscle in the form of small veins or venules, which replenish the supply to the abandoned area.

When one of the Discriminants is blocked, it also leads to dysfunction of the body's limbs: paralysis, cardiovascular accidents, synovitis, and many others. Even the human brain relies on a large number of discriminants for its blood and oxygen supply.

Discriminants give the heart the strength and capacity to pump blood well, until it circulates at its efficient speed. At this efficient speed, the body acquires the potential for immunocompetence. In other words, the organism has the capacity to repel any external or foreign agent. This is the body's natural defense.

We've just understood that muscles receive and register all impulses from external agents (forces), and this modifies the muscular surfaces, which become undulated, resulting in convulsions and tremors. So the Discriminant is an Arithmetico-geometric entity. For its application is Arithmetic and its demonstration is geometrical in a Grand-Carré of the Sacred Triangle of Intelligence containing two vertices, transcending and ascending. Where the light of the spirit travels.

That's why we're proposing the use of the Discriminant method in the global health system to cleanse the human skeleton and genome using natural light and digital pressure to reduce the mortality rate linked to a dysfunctional human skeleton suffocated by a lack of oxygen.

We're in the mathematical field of medical engineering here, and we need to define the concept properly to guide science in this new millennium specialty of technicians who clean or resuscitate the human skeleton and genome using natural light and digital pressure to advance humanity through research, innovation and discovery in time and space.

The Discriminant method is one solution among many to effectively address the many flaws and limitations of the global health system, which lacks the technicians to clean up the human skeleton and genome on a planetary scale.

Most of our illnesses come not only from our diet, but above all from the suffocation of the human skeleton, which no longer receives enough oxygen to function properly. Because oxygen is life. A public health problem whose mortality rate is impressive on a planetary scale.

This research thesis in the mathematics of medical engineering is a highly remarkable plea to the progress of science and humanity, which the WHO, UNESCO and all the nations of the world should seize upon to improve the global health system at the dawn of the third millennium.

The discriminant technique or method has already proved its worth on a small (local) scale, and needs to be generalized on a planetary scale. It's a question of the neurons of touch, in their remarkable role as detectors of the different types of balls by natural light and by digital pressure, which are at the root of a dysfunction of the human skeleton and genome.

Our fingers emit Lightwave energetic vibrations in the cleansing of the human skeleton and genome by light and digital pressure in a Grand-Carré, adjoining the fluid mechanisms; hence $F=p/s$.

Chapter III: What are the causes of Human Skeletal and Genome Dysfunction that the Discriminant Method will resolve?

Man is a fruit of interdisciplinarity whose essence of existence is mathematical as the roots of solar humanity, where we are its memory in the universality of the light of reason through a refinement of thought.

To each inheritance received from humanity's elders, add a fortune: this is the principle of continuity in discontinuity, to which we must all contribute in order to move humanity forward in its multiple problems of planetary dimensions, for which solutions must be found on the basis of achievements and contributions on a global scale.

The human body is a veritable crossroads of give and take, where there's plenty to eat and drink. It's like the Internet or social networks. A veritable selection garbage can for the good and the bad.

The human body is neither holy nor perfect. It is the fruit of an imperfection that requires updating. To better communicate with it, to get to know it better and discover its imperfections, in the quest for excellence in intellectual coherence or in the marriage of innovation and creation in a new frame of reference.

Poor diet and, above all, polluted air are the main causes affecting the human body, adjoining the individual's environment, creating mechanisms that combat the human body's defensive system, forming balls into grains of sand, kernels of corn and shards..., are responsible for a dysfunctional skeleton and human genome that suffocates when it no longer receives enough oxygen.

The skeleton becomes progressively deformed when these balls are not cleaned by natural light and digital pressure. This is also the case with sickle-cell patients, where the red blood cells deform into crescents or sickles to stifle the growth of the human skeleton and, above all, the proper circulation of oxygen.

These crescents sometimes end up taking on the nature of balls, grains of sand, corn kernels and shards that cause respiratory insufficiency and intense pain in the organism of sickle-cell subjects and many others, who need to be gradually cleansed by natural light and digital pressure to avoid this pain, and above all the deformation of the human skeleton due to the distension or spreading of the vertebrae.

Chapter IV: The Different Types of Balls that Cause Human Skeletal and Genome Dysfunction

Africa is the future of the science of the millennium through the revelation of the Zebra and Refracted Lightwave, based on intuition and mathematical astuteness through a refinement of thought.

We are in a world where commitment goes beyond the universality of the light of reason in its multiple problems of planetary dimensions, for which we are called to find solutions or make contributions on a global scale in the light of the wisdom of love.

Sand balls are most likely to be found in the rib cage, particularly in the ribs and intercostal muscles. It's because of their small size that they're known as sand-grain balls.

They are also found in the forehead, where they cause migraines, and in the sinuses, where they cause sinusitis. They are also found in the gums, where they form stones called tartars, causing the gums to bleed continuously.

In Africa in general, and in Congo Brazzaville in particular, the solution is to use firewood ash in the mouth for a maximum of two minutes. Modern medicine uses scaling. Sand balls are also at the root of bad breath.

These balls cause fevers that even the most high-performance or ultra-modern equipment can't detect these days. These balls are also found in the peritoneum, where they cause mild abdominal pain and constipation.

On the chin, these balls cause deformation of the mouth (facial paralysis). Sand-grain balls are the most frequent occurrence in sickle-cell patients. In this case, there are large numbers of balls at the ribs and at the base of the neck, where they prevent the blood supply to the head.

These balls cause severe respiratory insufficiency, manifested in sickle cell patients by jerky breathing.

Corn balls are large balls like corn kernels. They can be medium-sized or small, like all balls. They are responsible for poor blood circulation and are also capable of suffocating the human skeleton, which ends up with clogged pores. In other words, the skeleton becomes unable to send the red blood cells that become deformed to the bone marrow to supply the muscles with oxygen.

These balls in the lumbar region cause lumbago. In the knees, these balls cause synovitis and poor contact between joint surfaces. As a result, the individual topples to the ground. In the thoracic cavity, these balls are responsible for hypertension.

In the digestive tract, these balls are responsible for colopathy (malfunctioning of the intestines), and have no specific place in the human organism. They can be found everywhere.

Shard balls: these are balls that form after injection of oily products such as quinine, quinimax, etc., at a temperature of 37 degrees. This means that the product, which has not been fully eliminated by the body, crystallizes or turns into crystals and lodges in the muscles. These crystals are able to prick the patient's muscles internally.

Generally speaking, all balls suffocate the human skeleton, gradually deforming it. They are capable of distancing vertebrae from one another. The result is a deformed human skeleton.

All you need to do is clean them well, and they'll return to their normal position. The presence of a good quantity and quality of oxygen in the body is the fundamental principle of life. Indeed, there can be no life without oxygen.

Cleansing is therefore a resuscitator par excellence of human life, directly using aerobic dioxygen (oxygen from natural air), through the pressure of digital light applied to the surface of an organ. This is the discriminant technique or method applied in medical engineering.

Cleaning the human skeleton also helps unclog blood vessels (arteries, veins and capillaries), including the heart. If the latter has clots. It also cleans blood vessels in the hemispheres of the brain without opening the skull.

Using fluid mechanisms. That is, $F=p/s$ in a Grand-Carré. The fluid used is blood. This practice is demonstrated by the corkscrew experiment. We use digital pressure to cleanse the kidneys, nerves, red and white muscles, skeleton, knee, red and white blood, not forgetting chromosomes and other organs of the human body.

Its origins are mainly dietary, aerobic and environmental. The discriminant technique or method enables the organism to reinforce its natural immunity, conferring immunocompetent potential; the blood is said to acquire its effective speed. However, the body is not receptive to pathogens returning from elsewhere.

It has the ability to force them out. This is demonstrated by the experiment with an empty plastic bottle in a pond. This new knowledge shows that what remains to be discovered is greater than what is known today.

Every second that passes corresponds to a human discovery. A new knowledge is a madness of energy capable of provoking numerous accidents in certain intelligences; which do not see things in the same frame of reference as the scientist, the researcher or the genius in the combat of our time which is the essence and the urgency in the universality of the light of reason.

Chapter V: Energy variations in the human body: a global public health issue

The human body is a vast site of research and discovery in the universality of the light of reason, through a refinement of thought that is reflected on the screen of space-time.

The human body is a dynamic, a static being in the midst of an internal and external revolution that overturns everything in the metamorphoses of physical and natural metaphors into the mathematics of medical engineering, where research is a spirituality of solar souls that reside in the secrets and mysteries of the living pyramids contained within man himself as assets that must be preserved in the school of life.

Muscles receive controlled information containing energy. They have a memory and are able to record and retain the intensities of internal or external shocks received. These intensities are capable of modifying the muscular structure or surface, forming undulations.

These can emit repeated, high-intensity contractions that can tire or numb the nervous system. These involuntary movements can plunge the subject into a state of convulsion or coma.

Normally, energy should be evenly distributed throughout the human body, with no accumulation at any one point. But poor blood circulation at low speed, and poor muscular structuring, can make the muscle too rigid, sometimes even contracted without release, to the point of causing a variation in energy in the human body.

This translates into excruciating muscular pain in the human body or in isolated parts of the body. When an external force acts on any point in such a case, the point affected by the external force feels excruciating pain. This excess of energy at one point results in a deficit of energy at another point of the same human organism, called energy variation.

This variation in energy in the human body is characterized by muscular paralysis, a dysfunction that diminishes muscular joint movement, but with excruciating muscular pain. This energy variation can block arm-lifting and leg-bending movements, with a peak in the buttocks on one side towards the hip.

This variation in energy can be observed in the case of changes in muscle structure. That is, instead of being smooth, muscles may be rippled at their surfaces. This usually leads to repeated muscular contractions, resulting in convulsions and the subject falling into syncope or coma. This is because muscles have the power to register and accumulate energy from the shock of external impulses, however large or small.

The muscular response to these impulses may be delayed or sudden. This muscular state can be recognized by simply touching the muscle. The discriminant technique or method is the safest and most effective way of correcting these types of disorders, with muscle cleansing or re-education sessions using natural light and digital pressure to restore the muscles to their normal state.

The number of sessions required to correct such muscles varies from one subject to another, and also depends on the age of the subject or patient. Weak blood flow may be linked to blood coagulation, caused by elements in the external environment, either pathogenic agents (microbes) or poor-quality muscular structures, not forgetting balls of any kind, which are at the root of poor blood circulation in the human body.

The human skeleton is a large mosaic of discriminants that needs to be cleaned using natural light and digital pressure to circulate oxygen throughout the human body to clean all the organs: a public health problem with an impressive mortality rate due to a lack of skeletal and human genome cleaning technicians on a planetary scale.

Africa is providing a solution, or even a contribution, to this global problem. The health system, like the education system, is also one of the building blocks of nation-building. The two systems are linked. They are interdependent.

So WHO and UNESCO, not to mention humanity, must take this research thesis in Mathematics of Medical Engineering seriously. It's one of the paths to salvation for improving the world's health system, using the technique or discriminant method for cleaning the human skeleton and genome by natural light and digital pressure.

We need to revolutionize the global health system by introducing the mathematics of medical engineering to clean up the human skeleton and genome using the Discriminant method, which is simply the square of the difference between two numbers that are zero solutions of the second-degree trinomial.

Chapter VI: The world of medical engineering

The book of nature is the seat of energy, memory and intelligence. The intelligible universe can only be tamed by obeying its rites and customs, in accordance with the laws of nature and the universal light of reason.

Every innovation or invention is inspired by the book of nature, written in the language of nature's pure fractal mathematics.

It's through Grand-Amour and scientific reasoning that we show the power of discovery in medical engineering mathematics in the enigma of the destiny of space-time. In medical engineering, research means faithfully imitating nature in its authentic configuration, reflecting natural analogies.

The world of medical engineering is a refinement of ordinary thought that evolves through epistemological cuts from one research to another in a precise space-time.

We can see that the world of geometry and reified quantity has created a universe in which man no longer has a place. This universe is dominated by a symbolic language that erects a kind of wall between people, whose emotions are replaced by rationality, making communication between these two closely related concepts virtually impossible.

Whatever one man can imagine through the power of his thought, others are able to realize in the universal light of reason, through a refinement of mathematical thinking that is reflected on the screen of space-time.

For we live in a world of achievements and contributions to humanity's progress in intellectual production and self-education, where human intelligence escapes from one generation to the next, making its mark on history through intergenerational fire.

The existence of econometrics, sociometrics and biometrics, for example, testifies to the effort of human genius to use mathematics as the language of nature, in accordance with a tradition that has its roots in Pythagorean philosophy, which emerged from the Egyptian matrix: "everything is number" is the daughter of "the rule for studying nature; for understanding everything that exists, all the secrets, all the mysteries".

This hegemonic expression of reality, strongly embodied in the rigor of Galileo-Cartesian mathematics, unfairly excludes other forms of knowledge that paint the universe in which we live and love.

This world wonderfully woven by poetry and its universal music sharpens our sensitivity and feeds our unbridled imagination, generating inventive and creative geniuses through the revelation of the Zebra and Refracted Lightwave, based on intuition and mathematical astuteness, in a permanent dialogue between nature and mankind, whose spirituality is the only door of entry and exit into culture, science, art and the spiritual.

The Renaissance of Romanticism is an indispensable logical outcome of the desire to fill the existential void created by analytical geometry, which excludes the verbal model from the richer, more human reality considered to be the ideal path for the power of thought to understand reality.

The discovery of a mosaic of Discriminants in the composition of the human body is a great revolution in the Mathematics of Medical Engineering for the Cleaning of the Human Skeleton and Genome by natural light and digital pressure in a Grand-Carré, adjoining the fluid mechanisms. Hence $F=p/s$.

Every innovation or invention is the materialization of the Light Wave in exoteric and moral spirituality, expressed in the culture of a people in mathematical relation to real need.

The seeker is a lover of science and reason through a refinement of thought in the light of the wisdom of love.

Illustrations of Discriminants in the composition of the architectural framework of the human body.

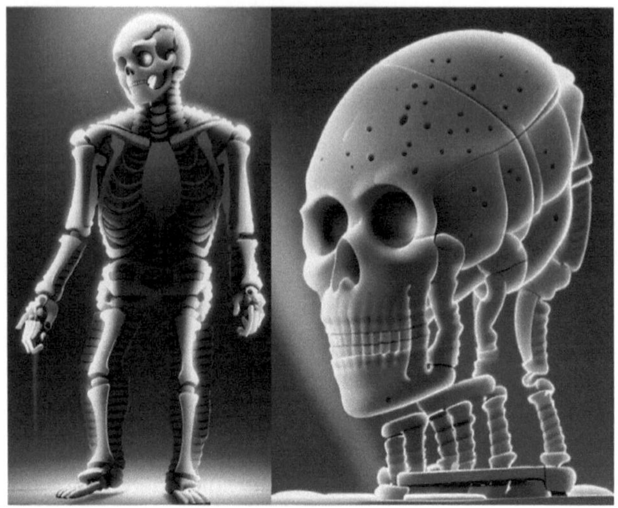

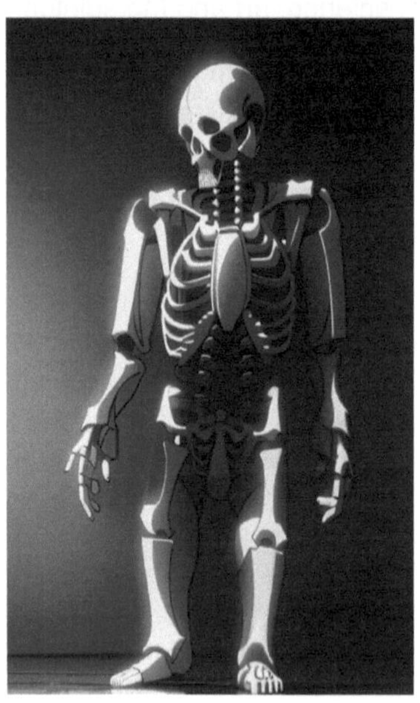

Conclusion

Elevated minds know neither hatred nor intellectual obscurantism, still less the apostasy of knowledge that circumscribes man's intelligence in dogmas and doctrines to be limited. They only know how to love what is true in the highlighting of true truth that reflects natural analogies, and to seek excellence as the search for truth in intellectual coherence and in the marriage of innovation and creation.

High spirits are always guided by the light of the wisdom of the love of humanity's elders, where the other cannot but be first in the universality of the light of reason. To this end, my thesis work in the mathematics of medical engineering is naturally one of many contributions to the advancement of humanity in its multiple problems of planetary dimensions, for which solutions must be found in the marriage of innovation and creation; in the totality of knowledge and in the knowledge of the totality in a twelve-dimensional space-time.

These days, modern science, education, public health and culture are going through a major crisis on a planetary scale, with all the nations of the world at a crossroads. Where will help come from?

Our rescue is only possible through research and innovation, invention and creativity, spirituality and discovery, interdisciplinarity, multidisciplinarity and transdisciplinarity, all essential assets for accessing true knowledge in the light of the spirit, through the refinement of thought.

African universities need to produce genuine research theses that contribute to solving global problems through research, innovation and discovery to meet the challenges of our time.

Africa doesn't need research outsourced to libraries on a copy-and-paste basis, without the slightest contribution or achievement in the field of science and literature.

True research doesn't mean copying and pasting from what already exists, without going beyond conventional analysis to create something new in an original and authentic way, in continuity and discontinuity. Because what already exists only serves as a reference point, insofar as we always build something new on the basis of what naturally exists.

So scientific research is not a compilation or collection of writings gathered in a library and then assembled around a subject with numerous pages with no effect of intellectual production in self-education leading to new knowledge that contributes to the progress of humanity in the universality of the light of reason.

Scientific research is a matter of achievements and contributions in innovation and creation of the mind through a refinement of mathematical thinking that is reflected on the screen of space-time in the cultural, scientific, artistic and spiritual.

When the future is bleak, the difficult becomes the path, and brilliant minds in the shadows reveal themselves in the struggle of their time to make history in the memory of generations. These are the challenges that enable man to question himself, to make a fresh start in managing the most urgent problems in the struggle of his time, through know-how that goes hand in hand with know-how to make his contribution to the heritage received from the elders of solar humanity.

In the dark cone, the scientist is the man who is always questioning himself to make a fresh start in the dialectic of the light of spirit and matter in motion to knot with his light cone for the search for truth in the questioning of established evidences as eternal truths, which must be redefined in true speech in a reality of true truth drawn from experimental soil.

Each level of knowledge has its own category of people. New knowledge is always misunderstood and arouses astonishment to such an extent that the researcher or scientist is even called crazy for being so far ahead of everyone else.

He's in a completely different frame of reference, and doesn't perceive the world in the same way as others. He's like a five- or seven-year-old child who perceives in his imagination a wonderful and sometimes terrifying world that others his age can't understand.

This is the spirit of the geniuses and scientists who are always misunderstood and sometimes rejected in their quest to find the real truth from experimental soil. True Scientific Research is the great love in itself for humanity in the light of the spirit.

The researcher or scholar never circumscribes his intelligence in dogmas that are like eternal truths to block the advancement of humanity in an apostasy of knowledge. He believes in dynamics, statics, innovation and the creation of the mind. For everything moves and dies by intergenerational fire.

The New African Dynamics refuse to be buried in humor in order to exist behind the march of history, where they will have neither contributions nor achievements to present at the crossroads of giving and receiving. We are an Africa of humanitude in culture, science, art and spirituality, revealed in the struggle of generations to pose the real problems of existentialism in the memory of solar humanity.

We are that memory of humanity which believes in the totality of knowledge and in the knowledge of the totality in the universality of the light of reason through a refinement of mathematical thought.

That's why you don't need to write an infinite number of pages to methodically prove the truth of a scientific fact. What is true can be demonstrated simply and clearly, without the slightest contradiction in the coherence of the scientific approach, by refining mathematical thinking in the light of the mind.

Afterword

The process of organ ageing begins at birth, but goes unnoticed by the general human consciousness, even though everything in life ages.

Nature's response to aging is self-maintenance, or channels such as discriminants for cleaning.

Professor Audrey KIBAMBA's thesis is revolutionary, and a great opportunity for humanity to realize that ageing is not just physical and superficial, but internal and profound, at the level of the vital organs.

Secondly, the thesis by research professor Audrey KIBAMBA shows man how to maintain a vital organ using the intelligence of neurons and the existence of discriminants designed for this purpose.

Professor Audrey KIBAMBA has shared one of the greatest secrets of longevity with humanity, free of charge.

In fact, premature death and longevity have been fighting an epic battle since birth, like a boxing match in the ring. Of the two, who will have the upper hand, who will inflict the ko on the other?

It would therefore seem salutary to vote for longevity by espousing Professor Audrey KIBAMBA's thesis, to give longevity a better chance of prevailing over premature death.

This thesis shows that living a long life is not a foregone conclusion, but the result of the cult of cleanliness in one's life, in the depths of the organs, by cleaning them very often using the discriminant method.

This thesis is therefore an invitation to longevity, which should be embraced and shared in every respect.

Medicine, in its drive to treat sick people, should incorporate Professor Audrey KIBAMBA's discriminant method into its pedagogy to raise its level of effectiveness.

This thesis should therefore be taught in medicine, to increase the effectiveness of doctors in dealing with diseases that are extremely difficult to diagnose.

Isn't the cause of this ailment the consequence of an organ with poor blood supply?

Professor Audrey KIBAMBA's thesis will answer all your questions.

Professor Audrey KIBAMBA is an opportunity for the world. He is the pride of Africa with his numerous theses published in Düsseldorf, Germany.

Thanks to Professor Audrey KIBAMBA for producing this thesis on behalf of humanity.

Only love of neighbor can inspire it.

Professor Paul YOTCHO, Mathematician, Physicist, Researcher, Writer

References

❖ Africa, Memory of Humanity, whose Essence and Urgency are the Combat of our time in a twelve-dimensional space-time. Author Audrey Kibamba de Bouansa gare de lumière, Editions Universitaires Européennes.

❖ Nouvelles Politiques de Développement des Nations Africaines dans un espace-temps à douze dimensions. Author Audrey Kibamba de Bouansa gare de lumière, Editions Universitaires Européennes.

❖ Knowledge and Humanity: Real Diplomas in Intellectual Production and Self-Education. Author Audrey Kibamba de Bouansa gare de lumière, Editions Universitaires Européennes.

Scientific Publications.

✓ Comment je vois les Belles Lettres: Théorie de la Poésie de l'Imagerie Scientifique dans un espace-temps à douze dimensions, published by le Lys Bleu Paris France in 2021. Author Audrey Kibamba de Bouansa gare de lumière.

✓ The Theory of the Twelve-Dimensional Universe on the Screen of Space-Time, published by European University Publishing in Düsseldorf, Germany, in 2021. Author Audrey Kibamba de Bouansa gare de lumière.

✓ Nouvelles Politiques de Développement des Nations Africaines dans un espace-temps à douze dimensions, published by European University Publishing in Düsseldorf,

Germany, in 2021. Author Audrey Kibamba de Bouansa gare de lumière.

✓ L'Afrique, Mémoire de l'Humanitude whose Essence and Urgency are the struggle of our time in a twelve-dimensional space-time, published by European University Publishing in Düsseldorf, Germany, in 2021. Author Audrey Kibamba de Bouansa gare de lumière.

✓ Connaissance et Humanité: les vrais diplômes dans la production intellectuelle dans l'auto-éducation, published by European University Publishing in Düsseldorf, Germany, in 2021. Author Audrey Kibamba de Bouansa gare de lumière.

✓ Qu'est-ce que la Poésie d'Equations Mathématiques dans le mariage de l'innovation et la création dans un espace-temps à douze dimensions, published by European University Publishing in Düsseldorf, Germany, in 2021. Author Audrey Kibamba de Bouansa gare de lumière.

✓ Positioning the Spatio-Temporal Continuum in the first fundamental relationship of surveying, published by European University Publishing in Germany in Düsseldorf in 2021. Author Audrey Kibamba de Bouansa gare de lumière.

✓ Discours sur la Théorie de la Poésie de l'arpentage lyrique dans un espace-temps à douze dimensions, published by European University Publishing in Germany in Düsseldorf in 2021. Author Audrey Kibamba de Bouansa gare de lumière.

✓ Les Douze Clés de la Connaissance sur l'Onde Lumineuse avec les lunettes de la Poésie Numérique et Quantique dans un Grand Carré, published by European University

Publishing in Düsseldorf, Germany, in 2021. Author Audrey Kibamba de Bouansa gare de lumière.

✓ Initiation à l'Epistémologie Numérique et Quantique dans un espace-temps à douze dimensions, published by European University Publishing in Düsseldorf, Germany, in 2022. Author Audrey Kibamba de Bouansa gare de lumière.

✓ Le Bloc Fédéral Panafricain Mécanismes des Réformes du Système Educatif Africain dans les Lumières des Humanités Classiques Africaines, published by European University Publishing in Germany in Düsseldorf in 2022. Author Audrey Kibamba de Bouansa gare de lumière.

✓ Denis Sassou N'Guesso at the heart of the challenges of deep African wisdom diplomacy in a multipolar world of balance and peace, published by European University Publishing in Germany in Düsseldorf in 2022. Author Audrey Kibamba de Bouansa gare de lumière.

✓ The Correlation between Science, Spirituality and Art in Research and Development, published by European University Publishing in Düsseldorf, Germany, in 2024. Author Audrey Kibamba de Bouansa gare de lumière.

✓ Introduction to the Twelve-Dimensional Theory of the Universe, published by Editions Universitaires Européennes in Düsseldorf, Germany, in 2024. Author Audrey Kibamba de Bouansa gare de lumière.

✓ Le Nettoyage du Squelette et du Génome Humain par la Méthode du Discriminant dans un Grand-Carré, published by Editions Universitaires Européennes in Düsseldorf, Germany, in 2024. Author Audrey Kibamba de Bouansa gare de lumière.

- ✓ L'arpentage lyrique, published by Muse in Düsseldorf, Germany, in 2021. Author Audrey Kibamba de Bouansa gare de lumière.

- ✓ L'aube des chants d'initiés, published by Editions le Lys Bleu Paris France in 2020. Author Audrey Kibamba de Bouansa gare de lumière.

Conference debates

- ✓ On the theme of Quantum Thinking at the heart of innovation. Author Audrey Kibamba de Bouansa gare de lumière on Mobali Makassi's youtube channel on February 19, 2023 at 7 p.m. Paris time, France.

- ✓

  On the theme Digital and Quantum Epistemology the key to Research and Development of African Nations. Author Audrey Kibamba de Bouansa gare de lumière sur la chaine youtube Causons d'Afrique le 17 novembre 2022 à 20 heures heure de Paris France.

- ✓ On the theme Cleaning the Human Skeleton and Genome with Natural Light and Digital Pressure. Author Audrey Kibamba de Bouansa gare de lumière sur la voix de la diaspora on June 7, 2022 at 3 p.m. Abidjan time in Côte d'Ivoire.

- ✓ On the theme How I see the Belles Lettres in a twelve-dimensional space-time. Author Audrey Kibamba de Bouansa gare de lumière sur la chaine youtube huit milles tambours d'Afrique le 20 Mars 2022 à 17 heures heure de Bruxelles en Belgique.

- ✓ On the theme of the Correlation between Science, Spirituality and Art in Research and Development; and the Theory of the Twelve-Dimensional Universe. Author Audrey Kibamba de Bouansa light station at the Faculté des Sciences et Techniques de l'Université Marien Ngouabi on February 20, 2024 at 10 a.m. Brazzaville time.

I want morebooks!

Buy your books fast and straightforward online - at one of world's fastest growing online book stores! Environmentally sound due to Print-on-Demand technologies.

Buy your books online at
www.morebooks.shop

Kaufen Sie Ihre Bücher schnell und unkompliziert online – auf einer der am schnellsten wachsenden Buchhandelsplattformen weltweit! Dank Print-On-Demand umwelt- und ressourcenschonend produzi ert.

Bücher schneller online kaufen
www.morebooks.shop

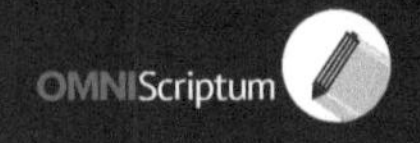

Printed by Books on Demand GmbH, Norderstedt / Germany